ISBN 978-1-5278-9327-6
PIBN 10926501

1 MONTH OF
FREE
READING

at

www.ForgottenBooks.com

By purchasing this book you are eligible for one month membership to ForgottenBooks.com, giving you unlimited access to our entire collection of over 1,000,000 titles via our web site and mobile apps.

To claim your free month visit:

www.forgottenbooks.com/free926501

English
Français
Deutsche
Italiano
Español
Português

www.forgottenbooks.com

Mythology Photography **Fiction**
Fishing Christianity **Art** Cooking
Essays Buddhism Freemasonry
Medicine **Biology** Music **Ancient**
Egypt Evolution Carpentry Physics
Dance Geology **Mathematics** Fitness
Shakespeare **Folklore** Yoga Marketing
Confidence Immortality Biographies
Poetry **Psychology** Witchcraft
Electronics Chemistry History **Law**
Accounting **Philosophy** Anthropology
Alchemy Drama Quantum Mechanics
Atheism Sexual Health **Ancient History**
Entrepreneurship Languages Sport
Paleontology Needlework Islam
Metaphysics Investment Archaeology
Parenting Statistics Criminology
Motivational

CIHM/ICMH Microfiche Series.

CIHM/ICMH Collection de microfiches.

ute for Historical Microreproductions / Institut canadien de microreproductior

198

The Institute has attempted to obtain the best original copy available for filming. Features of this copy which may be bibliographically unique, which may alter any of the images in the reproduction, or which may significantly change the usual method of filming, are checked below.

L'Institut a
qu'il lui a é
de cet exe
point de v
une image
modificati
sont indiqu

☑ Coloured covers/
Couverture de couleur

☐ Covers damaged/
Couverture endommagée

☐ Covers restored and/or laminated/
Couverture restaurée et/ou pelliculée

☐ Cover title missing/
Le titre de couverture manque

☐ Coloured maps/
Cartes géographiques en couleur

☐ Coloured ink (i.e. other than blue or black)/
Encre de couleur (i.e. autre que bleue ou noire)

☐ Coloured plates and/or illustrations/
Planches et/ou illustrations en couleur

☐ Bound with other material/
Relié avec d'autres documents

☐ Tight binding may cause shadows or distortion along interior margin/
La reliure serrée peut causer de l'ombre ou de la distorsion le long de la marge intérieure

☐ Blank leaves added during restoration may appear within the text. Whenever possible, these have been omitted from filming/
Il se peut que certaines pages blanches ajoutées lors d'une restauration apparaissent dans le texte, mais, lorsque cela était possible, ces pages n'ont pas été filmées.

☐ Additional comments:/
Commentaires supplémentaires:

☐ Colo
Page

☐ Page
Page

☐ Page
Pages

☑ Pages
Pages

☐ Pages
Pages

☑ Show
Trans

☐ Quali
Quali

☐ Inclu
Comp

☐ Only
Seule

☐ Pages
slips,
ansur
Les p
obscu
etc.,
obten

n reproduced thanks

L'exemplaire filmé fut reproduit grâce à la générosité de:

Bibliothèque,
Commission Géologique du Canada

e the best quality
ition and legibility
eping with the
s. .

Les images suivantes ont été reproduites avec l
plus grand soin, compte tenu de la condition et
de la netteté de l'exemplaire filmé, et en
conformité avec les conditions du contrat de
filmage.

er covers are filmed
r and ending on
r illustrated impres-
appropriate. All
beginning on the
ustrated impres-
age with a printed

Les exemplaires originaux dont la couverture en
papier est imprimée sont filmés en commençan
par le premier plat et en terminant soit par la
dernière page qui comporte une empreinte
d'impression ou d'illustration, soit par le second
plat, selon le cas. Tous les autres exemplaires
originaux sont filmés en commençant par la
première page qui comporte une empreinte
d'impression ou d'illustration et en terminant p
la dernière page qui comporte une telle
empreinte.

ich microfiche
(meaning "CON-
meaning "END"),

Un des symboles suivants apparaîtra sur la
dernière image de chaque microfiche, selon le
cas: le symbole ⟶ signifie "A SUIVRE", le
symbole ▽ signifie "FIN".

ly be filmed at
se too large to be
ure are filmed
nd corner, left to
any frames as
ms illustrate the

Les cartes, planches, tableaux, etc., peuvent êt
filmés à des taux de réduction différents.
Lorsque le document est trop grand pour être
reproduit en un seul cliché, il est filmé à partir
de l'angle supérieur gauche, de gauche à droite
et de haut en bas, en prenant le nombre
d'images nécessaire. Les diagrammes suivants
illustrent la méthode.

2   3

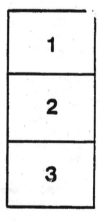

1   2   3

ON

THE LAURENTIAN ROCKS

OF

BAVARIA.

BY DR. GÜMBEL.

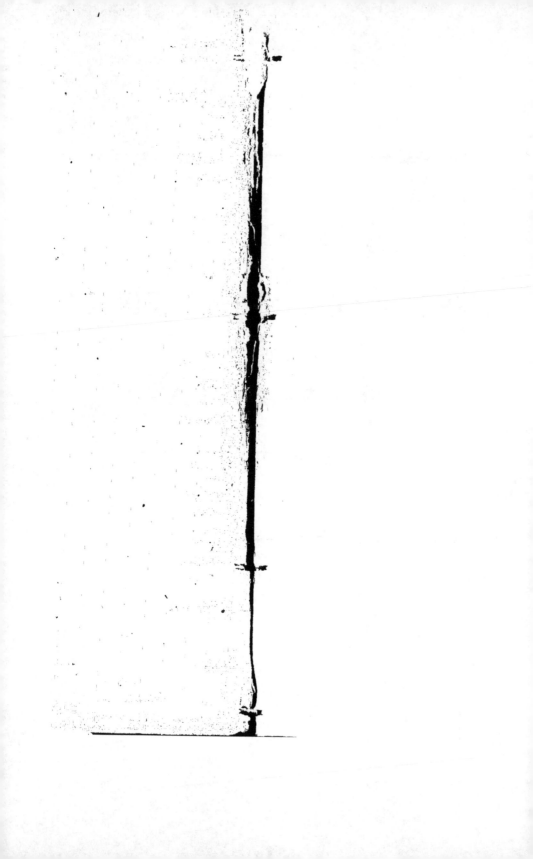

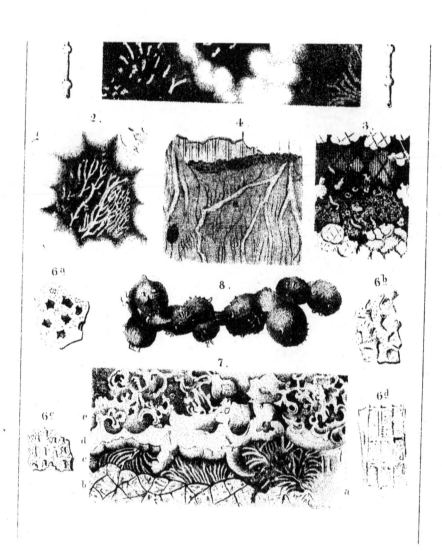

LAURENTIAN ROCKS OF BAVARIA

By Dr. Gümbel,

Director of the Geological Survey of Bavaria; with a plate containing figures of two species of Eozoon.

(*From the* Canadian Naturalist *for December*, 1866.)

———

The discovery of organic remains in the crystalline limestones of the ancient gneiss of Canada, for which we are indebted to the researches of Sir William Logan and his colleagues, and to the careful microscopic investigations of Drs. Dawson and Carpenter, must be regarded as opening a new era in geological science.

This discovery overturns at once the notions hitherto commonly entertained with regard to the origin of the stratified primary limestones, and their accompanying gneissic and quartzose strata, included under the general name of primitive crystalline schists. It shows us that these crystalline stratified rocks, of the so-called primary system, are only a backward prolongation of the chain of fossiliferous strata ; the elements of which were deposited as oceanic sediment, like the clay-slates, limestones and sandstones of the paleozoic formations, and under similar conditions, though at a time far more remote, and more favorable to the generation of crystalline mineral compounds.

In this discovery of organic remains in the primary rocks, we hail with joy the dawn of a new epoch in the critical history of these earlier formations. Already, in its light, the primeval geologic time is seen to be everywhere animated, and peopled with new animal forms, of whose very existence we had previously no suspicion. Life, which had hitherto been supposed to have first

—————————————————

*Editor's Note.—In revising and preparing this for the press, the original paper has been considerably abridged by the omission of portions, whose place is indicated in the text. Some explanatory notes have also been added.—T. S. H.

appeared in the primordial division of the Silurian period, is now seen to be immeasurably lengthened beyond its former limit, and to embrace in its domain the most ancient known portions of the earth's crust. It would almost seem as if organic life had been awakened simultaneously with the solidification of the earth's crust.

The great importance of this discovery cannot be clearly understood, unless we first consider the various and conflicting opinions and theories which had hitherto been maintained concerning the origin of these primary rocks. Thus some, who consider them as the first-formed crust of a previously molten globe, regard their apparent stratification as a kind of concentric parallel structure, developed in the progressive cooling of the mass from without. Others, while admitting a similar origin of these rocks, suppose their division into parallel layers to be due, like the lamination of clay-slates, to lateral pressure. If we admit such views, the igneous origin of schistose rocks becomes conceivable, and is in fact maintained by many.

On the other hand, we have the school which, while recognizing the sedimentary origin of these crystalline schists, supposes them to have metamorphosed at a later period; either by the internal heat, acting in the deeply buried strata; by the proximity of eruptive rocks; or finally, through the agency of permeating waters charged with certain mineral salts.

A few geologists only have hitherto inclined to the opinion that these crystalline schists, while possessing real stratification, and sedimentary in their origin, were formed at a period when the conditions were more favorable to the production of crystalline materials than at present. According to this view, the crystalline structure of these rocks is an original condition, and not one superinduced at a later period by metamorphosis. In order however to arrange and classify these ancient crystalline rocks, it becomes necessary to establish, by superposition or by other evidence, differences in age, such as are recognized in the more recent stratified deposits. The discovery of similar organic remains, occupying a determinate position in the stratification, in different and remote portions of these primitive rocks, furnishes a powerful argument in favor of the latter view, as opposed to the notion which maintains the metamorphic origin of the various minerals and rocks of these ancient formations; so that we may regard the direct formation of these mineral elements, at least so far as these fossiliferous primary limestones are concerned, as an established fact.

So early as 1853, after investigating the primitive rocks of eastern Bavaria, which are connected with those of the Bohemian forest, I expressed the opinion that, although eruptive masses of granite and similar rocks occur in that region,.the gneiss was of sedimentary origin, and divisible into several formations. I at that time endeavored to separate these crystalline schists into three great divisions, the phyllades, the mica-schists, and the gneiss formation, of which the first was the youngest and the last the oldest; all these formations having essentially the same dip and strike.

These results, obtained from very detailed geological and topographical researches, were subsequently more fully set forth in the Survey of the Geology of Eastern Bavaria, (Book IV., p. 219 *et seq.*); where I endeavored to assign local names to the subdivisions of the primitive rocks of that region. Beginning with the more recent, I distinguished the following formations :

1. Hercynian primitive clay-slate.
2. Hercynian mi⸱
3. Hercynian gn⸱   y gneiss system.
4. Bojian gneiss.

In some cases, within   as, I even succeeded in tracing
out still smaller subdivisi⸱   vas in this way established that
definite and distinct kinds of rocks, as for example hornblende-slate and mica-slate, may replace each other and, as it were, pass into each other, in different parts of the same horizon.

After Sir Roderick Murchison had established the existence of the fundamental gneiss in Scotland, and recognized its identity with that of the Laurentian system of Canada, he turned his attention to the primitive rocks of Bavaria and Bohemia. My researches and my communications to him disclosed the important fact that these rocks belong to the same series as the oldest formations of Canada and of Scotland. On one point only was there an apparent difference of opinion between Sir Roderick and myself; which was that he was disposed to look upon the whole of the gneiss of the Hercynian mountains as constituting but a single formation, corresponding to the Laurentian gneiss of Canada and of Scotland; while I had endeavored to distinguish two divisions, the newer grey or Hercynian gneiss, and the older red or variegated, which I called the Bojian gneiss. This difference of opinion is however at once removed by the remark that I did not intend to maintain in the older gneiss the existence of a formation more ancient than the fundamental gneiss of Scotland, nor yet to assimilate the newer or grey gneiss to the more recent or so-called metamorphic series, which, according to Sir Roderick, may be clearly distinguished in Scotland from the Laurentian gneiss.

[This newer gneissic formation of the Highands is, according to Murchison, Ramsay and others, of Lower Silurian age. Our author simply claims to have established a division in the proper Laurentian rocks of Bavaria and Bohemia. It will be seen from the recently published maps of the Laurentian region of the Ottawa, that Sir William Logan there distinguishes three great limestone formations, by which the enormous mass of Laurentian gneiss is separated into four divisions. One or two of the upper ones of these may be eventually found to correspond to the grey Hercynian gneiss of Bavaria, which is there accompanied by the Eozoon Canadense, a fossil so far as yet known characterizing the highest of the three Laurentian limestones. This grey gneiss of Bavaria appears to be lithologically distinct from the Labrador (or Upper Laurentian) series; nor do we find in the present memoir of Gumbel, any clear evidence of the occurrence either of this, or of the Huronian system, in Bavaria.—T. S. H.

After citing in this connection Sir W. E. Logan's observations on these ancient formations, which are shown, by the results of the Canadian Survey, to represent three great systems of sedimentary rocks, formed under conditions not unlike those of more modern formations, our author observes:—]

Accepting these views of the older Canadian rocks, it would naturally follow that organic life might be expected to reach back much farther than the so-called primordial fauna of Lower Silurian age, and to mark the period hitherto designated as Azoic.

Guided by these ideas, the geologists of Canada zealously sought for traces of organic life in the primitive rocks of that country. Dr. Sterry Hunt had already concluded that it must have existed in the Laurentian period, from the presence of beds of iron ore, and of metallic sulphurets, which, not less than the occurrence of graphite, were to him chemical evidences of an already existing vegetation, when at length direct evidence of life was obtained by the discovery of apparently organic forms in the great beds of crystalline limestone which occur in the Laurentian system. Such were collected in 1858, by Mr. J. McMullen from the Grand Calumet on the Ottawa River, and were observed by Sir Wm. Logan to resemble closely similar specimens obtained by Dr. James Wilson in Burgess, a few years previously. In 1859, Sir Wm. Logan first expressed his opinion that these masses, in which pyroxene, serpentine, and an allied mineral, alternated in thin layers, with carbonate of lime or dolomite, were of organic origin; and in 1862 he reiterated this opinion in England, without however being able to convince the English geologists, Ramsay excepted, of the correctness of his views. Soon after this, however, the discovery of other and more perfect specimens, at Grenville, furnished decisive proofs of the organic nature of these singular fossils.

The careful and admirable investigations of Dawson and of Carpenter, to whom specimens of the rock were confided, have placed beyond doubt the organic structure of these remains, and confirmed the important fact that these ancient Laurentian limestones abound in a peculiar organic fossil, unknown in more recent formations, to which has been given the name of Eozoon.*

＊ 　 　 ＊ 　 　 ＊ 　 　 ＊ 　 　 ＊ 　 　 ＊

The researches of Sterry Hunt on the mineralogical relations of the Eozoon-bearing rocks, lead him to the important conclusion that certain silicates, namely serpentine, white pyroxene, and loganite, have filled up the vacant spaces left by the disappearance of the destructible animal matter of the sarcode, the calcareous skeleton remaining more or less unchanged. If, by the aid of acids, we remove from such specimens the carbonate of lime, (or, in certain cases, the dolomite which replaces it,) there remains a coherent skeleton, which is evidently a cast of the soft parts of the Eozoon. The process by which the silicates have been introduced into the empty spaces corresponds evidently to that of ordinary silicification through the action of water. It is to be noted that Hunt found serpentine and pyroxene, side by side, in adjacent chambers, and even sharing the same chamber between them; thus affording a beautiful proof of their origin through the

---

* Here follows, in the original, a lengthened analysis of the memoirs of Messrs. Logan, Dawson, Carpenter, and Hunt, published in the Quarterly Journal of the Geological Society of London, and already reprinted in the Canadian Naturalist.

infiltration of aqueous solutions, while the Eozoon was yet growing, or shortly after its death.   *   *   *

Hunt, in a very ingenious manner, compares this formation and deposition of serpentine, pyroxene, and loganite, with that of glauconite, whose formation has gone on uninterruptedly from the Silurian to the Tertiary period, and is even now taking place in the depths of the sea; it being well known that Ehrenberg and others have already shown that many of the grains of glauconite are casts of the interior of foraminiferal shells.   In the light of this comparison, the notion that the serpentine, and such like minerals of the primitive limestones have been formed in a similar manner, in the chambers of Eozoic foraminifera, loses any traces of improbability which it might at first seem to possess.   *   *

My discovery of similar organic remains in the serpentine-limestone from near Passau was made in 1865, when I had returned from my geological labors of the summer, and received the recently published descriptions of Messrs. Logan, Dawson, etc. Small portions of this rock, gathered in the progress of the geological survey in 1854, and ever since preserved in my collection, having been submitted to microscopic examination, confirmed in the most brilliant manner the acute judgment of the Canadian geologists ; and furnished paleontological evidence that, notwithstanding the great distance which separates Canada from Bavaria, the equivalent primitive rocks of the two regions are characterized by similar organic remains; showing at the same time that the law governing the definite succession of organic life on the earth is maintained even in these most ancient formations. The fragments of serpentine-limestone or ophicalcite, in which I first detected the existence of Eozoon, were like those described in Canada in which the lamellar structure is wanting, and offer only what Dr. Carpenter has called an acervuline structure.   For further confirmation of my observations, I deemed it advisable, through the kindness of Sir Charles Lyell, to submit specimens of the Bavarian rock to the examination of that eminent authority, Dr. Carpenter ; who, without any hesitation, declared them to contain Eozoon.

This fact being established, I procured from the quarries near Passau as many specimens of the limestone as the advanced season of the year would permit ; and, aided by my diligent and skilful assistants Messrs. Reber and Schwager, examined them by the methods indicated by Messrs. Dawson and Carpenter.   In this way I soon convinced myself of the general similarity of our organic remains with those of Canada. Our examinations were made on polished sections and in portions etched with dilute nitric acid, or, better, with warm acetic acid. The most beautiful results were however obtained by etching moderately thin sections, so that the specimens may be examined at will either by reflected or by transmitted light.

The specimens in which I first detected Eozoon came from a quarry at Steinhag, near Obernzell on the Danube, not far from Passau. The crystalline limestone here forms a mass from fifty to seventy feet thick, divided into several beds, included in the gneiss, whose general strike in this region is N.W., with a dip of 40°–60° N.E. The limestone strata of Steinhag have a dip of 45° N.E.. The gneiss of this vicinity is chiefly grey, and very silicious, containing dichroite, and of the variety known as dichroite-gneiss; and I conceive it to belong, like the gneiss of Bodenmais and Arber, to that younger division of the primitive gneiss system which I have designated as the Hercynian gneiss formation; which both to the north, between Tischenreuth and Mahring, and to the south, on the south-west of the mountains of Ossa, is immediately overlaid by the mica-slate formation. Lithologically, this newer division of the gneiss is characterized by the predominance of a grey variety, rich in quartz, with black magnesian-mica and orthoclase, besides which a small quantity of oligoclase is never wanting. A farther characteristic of this Hercynian gneiss is the frequent intercalation of beds of rocks rich in hornblende, such as hornblende-schist, amphibolite, diorite, syenite, and syenitic granite, and also of serpentine and granulite. Beds of granular limestone, or of calcareous schists are also never altogether wanting; while iron pyrites, and graphite, in lenticular masses, or in local beds conformable to the great mass of the gneiss strata, are very generally present.

The Hercynian gneiss strata on the shores of the Danube near Passau are separated from the typical Hercynian gneiss districts which occur to the north, on the borders of the Fichtelgebirge and near Bodenmais and Arber, by an extensive tract, partly occupied by intrusive granites, and partly by another variety of gneiss. These Danubian gneiss strata are not seen to come in contact with any newer crystalline formation, but towards the south are concealed by the tertiary strata of the Danubian plain; while towards the N.W. they are in part cut off by granite, and in part replaced by those belts of gneiss which accompany the quartz ridge of the Pfahl, and belong to the red variety or Bojian gneiss. The grey gneiss strata of the Danube might therefore be supposed to be older than this red gneiss, which from its relations in the district to the N.W., between Cham and Weiden, I had regarded as itself the more ancient formation. But the lithological characters of the grey Danubian gneiss are opposed to this view, since this rock not only presents a general resemblance to the gneiss formation of Bodenmais, which without doubt is directly overlaid by the mica-schist of the mountains of Ossa, thus shewing it to be the newer gneiss; but exhibits a repetition of the minor features which characterize the gneiss district of Bodenmais. We find in the Danubian gneiss that same abundant dissemination of dichroite, which gives rise to the typical dichroite-gneiss of Bodenmais, with nearly the same mineral associations in both cases. On the Danube, also, interstratified beds of hornblende-rock (at Hals near Passau), of serpentine (at Steinhag), and of pyrites (at Kelberg, and many points along the Danube), occur, as in the north. On the other hand, the graphite which abounds in the gneiss of Passau is not wanting at Bodenmais or Tischenreuth. The interstratified syenites and syenitic granites are, in like manner, common to all these districts; those near Passau being, however, richer in easily decomposed minerals, such as porcelain-spar (scapolite) and calcspar, are more subject to decomposition, and form the parent rock of the famous porcelain clays of the region.

These resemblances lead me to refer the Danubian gneiss, notwithstanding its apparent stratigraphical inferiority to the red gneiss, to the newer or Hercynian formation; and to explain its apparently abnormal relations by assuming a fault running along the strike from N.W. to S.E., through which the older gneiss of the Pfahl is brought up, and seems to overlie the younger.

We shall then regard the whole of the gneissic strata character-ized by dichroite, which extend on the Danube from Passau to Linz, as equivalent to the Hercynian gneiss of Bodenmais, and designate it as the Danubian gneiss. We may here call attention to the abundance of graphitic beds in it, as also to the occurrence of porcelain clay, and of beds of iron pyrites and magnetic pyrites. If it is true (as maintained by Dr. Sterry Hunt) that all graphite owes its origin to organic matters, we must suppose the existence of a primordial region peculiarly rich in organic life; since graphite occurs here in almost all the strata, and in some places in

such quantities that it is profitably extracted, and is largely used
for the manufacture of the famous Passau crucibles.   In all of
the numerous graphite mines, the uniform interstratification of
bands and lenticular masses rich in graphite with the gneiss
is here distinctly marked.   A similar arrangement is seen in the
sulphurets of iron, which are more abundantly disseminated in the
more hornblendic strata.   The localities of porcelain-earth or
kaolin are in like manner confined to the strike of the gneissic
strata; and are generally contiguous to certain interstratified
granitic and syenitic bands, rich in feldspar.   Its frequent
association with porcelain-spar, (probably nothing more than
a chloriferous scapolite or anorthite,) indicates that this mineral
has played an essential part in the production of the kaolin.   The
presence of chlorine in this mineral is highly significant, and
suggests the agency of sea-water in its production.

Of particular interest, from their mineral associations, are three
or more parallel bands of crystalline limestone of no great
thickness, which occur conformably interstratified with the gneiss
of the hills near Passau.   They begin near Hofkirchen, and
extend north and south, from along the Danube as far as the
frontier, near Jochenstein, where the Danube leaves Bavaria.
These separate limestone bands, although exposed by numerous
quarries, cannot be followed uninterruptedly, being sometimes
concealed, and sometimes of insignificant thickness.

*   *   *   *   *   *

The large quarry of Steinhag already described, from which I
first obtained the Eozoon, is one.   The enclosing rock is a grey
hornblendic gneiss, which sometimes passes into a hornblende-
slate.   The limestone is in many places overlaid by a bed of
hornblende-schist, sometimes five feet in thickness, which separates
it from the normal gneiss.   In many localities, a bed of serpentine,
three or four feet thick, is interposed between the limestone and
the hornblende-schist; and in some cases a zone, consisting chiefly
of scapolite, crystalline and almost compact, with an admixture
however of hornblende and chlorite.   Below the serpentine band,
the crystalline limestone appears divided into distinct beds, and
encloses various accidental minerals, among which are reddish-
white mica, chlorite, hornblende, tremolite, chondrodite, rosellan,
garnet, and scapolite arranged in bands.   In several places the
lime is mingled with serpentine, grains or portions of which, often
of the size of peas, are scattered through the limestone with

apparent irregularity, giving rise to a beautiful variety of ophicalcite or serpentine-marble.   These portions, which are enclosed in
the limestone destitute of serpentine, always present a rounded
outline.   In one instance there appears, in a high naked wall of
limestone without serpentine, the outline of a mass of ophicalcite,
about sixteen feet long and twenty-five feet high, which, rising
from a broad base, ends in a point, and is separated from the
enclosing limestone by an undulating but clearly defined margin, as
already well described by Wineberger.   This mass of ophicalcite
recalls vividly a reef-like structure.   Within this, and similar
masses of ophicalcite in the crystalline limestone, there are, so far
as my observations in 1854 extend, no continuous lines or
concentric layers of serpentine to be observed, this mineral being
always distributed in small grains and patches.   The few
apparently regular layers which may be observed are soon
interrupted, and the whole aggregation is irregular.   [This is
well shown in plates II. and III. in the original memoir, which
recall the acervuline portions, that make up a large part of the
Canadian specimens of Eozoon.—EDS.]

The numerous specimens which were subsequently collected, at
the commencement of the winter, show, throughout, this irregular
structure, which seems to characterize the Bavarian specimens of
Eozoon, as is in part the case in those from Canada.   It is true
that small lenticular masses or nodules, consisting chiefly of
scapolite, measuring fifty by twenty millimeters, and even much
more, are often met with, around which serpentine is arranged in
a concentric manner ; but even here the serpentine is in small
cohering masses, and not in regular layers ; nor could I, after
numerous examinations of fragments of such masses, satisfy
myself whether I had to deal with the commencing growth of an
Eozoon, or merely with a concretionary mass; since the granular
structure of the scapolite centre could never be clearly made out.
Moreover the occurrence of these nodules, arranged in a stratiform
manner, is opposed to the notion that they are nuclei of Eozoon,
although in the parts around these nodules I could sometimes
distinctly observe tubuli, canals, and even indications of a shell-like
structure.

The portions of serpentine in the ophicalcite occur of very
various sizes, from that of a millet-seed to lumps whose sections
measure fifteen by six or eight millimeters.   But I think I can
detect within certain lines, (which are not, it is true, very well

defined,) chains of serpentine grains, of nearly equal size, connected with each other. When by means of acids the lime is removed from these aggregates, a perfectly coherent serpentine skeleton is in all cases obtained, which may be compared to a piece of wood perforated by ants.   *   *   *   *   *

The surface of the serpentine grains is rounded, pitted, and irregular; plane surfaces and straight lines are rarely to be seen. Even when dilute nitric or acetic acid has been used to remove the lime, a white down-like coating is frequently found on the serpentine, which does not answer to the nummuline wall of the calcareous skeleton. In many cases, where the lime is very crystalline, and the more delicate organic structure obliterated, small tufts of radiated crystals, apparently hornblende or tremolite, are seen resting upon the serpentine. These crystals, when seen in thin sections, by transmitted light, may easily give rise to errors; their formation seems to have been possible only where the calcareous skeleton had been destroyed, and crystalline carbonate of lime deposited in its stead; during which time free space was given for the formation of these crystalline groups. In very many cases there are seen, by a moderate magnifying power, (in the residue from acids) deposits of small detached cylindrical stems, with some larger ones, consisting of a white matter insoluble in acids. These appear to be the casts of the tubuli which penetrated the calcareous skeleton, and of the less frequent stolons, as will be described.

The serpentine in these sections never appears quite homogeneous, but exhibits, on the contrary, irregular groups of small dark-colored globules disseminated through the mass, without however any definite indications of organic form. Still more frequently, the serpentine is penetrated by irregularly reticulated dark colored veins, giving to the mass a cellular aspect.

In certain parts of the serpentine, however, parallel lines, groups of curved tube-like forms, and oval openings, clearly indicate an organic structure like that of the Canadian Eozoon. The finely tubulated nummuline wall of the chambers, which was discovered by Carpenter, and the casts of whose tubuli appear in the decalcified specimens from Canada as a soft white velvet-like covering, could only be found in a few isolated cases in the Bavarian specimens, but was clearly made out in a few fragments. (Pl. I., 4.) The somewhat oblique section shows the openings of the minute tubuli.

It should be remarked that the serpentine at Steinhag occurs, not only replacing the sarcode in the carbonate of lime of the Eozoon, but also forming layers over the limestone strata, and moreover filling up large and small crevices and fissures, which have nothing at all to do with the organic structure. Especially worthy of notice are the plates of fibrous serpentine, or chrysotile, often from five to ten millimeters in diameter, which are found extending in unbroken lines through the compact serpentine.

The color of the serpentine presents all possible shades, from blackish green, to the palest yellowish green tint. Where it has been exposed to the weather, the serpentine has become of a pale brownish green, and appears changed into gymnite. The different tints are arranged in zones, and seem to mark different periods of growth. The carbonate of lime which is interposed among the grains of serpentine in the specimens from Steinhag, is either distinctly crystalline, or apparently compact. In the first case, no organic structure can be perceived; thin sections of the crystalline portions show only intersecting parallel lines; and in etched or entirely decalcified specimens, no clear evidence of the fine canal-system of the skeleton can be observed. These crystalline portions often alternate with others which are compact and but feebly translucent. In thin sections of these compact parts, the rounded forms of the delicate tubuli are very clearly discerned, provided the section is at right angles to them. In etched specimens, viewed by reflected light, these tubuli are seen to branch out in the form of tufts, exactly as described and figured by Drs. Dawson and Carpenter.

These branching and ramified tubuli rest upon the serpentine granules, and seem by anastomosis to be connected with adjacent groups. The diameter of these tubuli is from $\frac{1}{100}$ to $\frac{3}{1000}$ millimeters. They are easily distinguishable from the delicate groups of crystals, which are also sometimes found implanted in the serpentine, by the nearly uniform thickness throughout their whole length; by their extremities, which are always somewhat crooked; and by their pipe-like form. The latter are never ramified; have a fibrous aspect; and are always straight, and terminate in a point. (Pl. I., figs. 1, 2, 3.)

Here and there are observed larger tubuli, which, so far as my observations extend, are always isolated, and nearly or quite parallel. (Pl. I., fig. 1.) Their diameter is about $\frac{1}{100}$ millimeters, and they not improbably represent those stolons or connecting channels with which Carpenter has made us acquainted.

In the decalcified specimens, delicate very slender string-like leaflets were very frequently observed, stretched between the serpentine granules; but they presented no discernible organic structure, and are perhaps only the casts of small crevices. More remarkable are the numerous canals filled with carbonate of lime, which traverse the serpentine granules, and at the surface of these are expanded into funnel shapes. They appear to represent cross connections between the calcareous skeleton.

As my object at present is merely to shew the presence, in the primitive limestones of Bavaria, of forms corresponding to the Canadian Eozoon, I will not dwell longer on these various appearances met-with in the microscopical examinations, nor on the peculiar cellular structures observed in the carbonate of lime. I will, for the same reason, only mention a specimen which exhibits, by the side of a curved main tube, a number of secondary tubuli, and farther on a parallel layer of fibres; and also another radiated form which resembles a section of a Bryozoon. It is sufficient to draw attention to the fact that, in addition to Eozoon, there are other organic remains in these crystalline limestones. There remains however to be noticed a phenomenon of some importance.

When the lime is removed by nitric or acetic acid from the interstices of the serpentine granules, there may be observed, on gently moving the liquid, extremely delicate membranes, that separate themselves from the serpentine grains, (which they covered thickly, as with a fine white down,) and now remain swimming in the liquid, so that they can readily be separated, by decantation, from a multitude of heavier particles, which, having also detached themselves from the serpentine mass, accumulate at the bottom of the vessel. These consist in great part of indistinct mineral fragments, and of small crystalline needles, together with distinct cylindrical portions, which are the broken tubuli of the Eozoon. Besides these are, here and there, distinctly knotted stems or tubules, (Pl. I., figs. 5, a and b,) which I dare not suppose to belong to Eozoon. Various other fragments of tubuli are also associated with these.

The delicate flakes, which can be obtained by evaporating the liquid in which they are suspended, shew, under a magnifying power of 400 diameters, a membranous character, and peculiar structures, which seem to be undoubtedly of organic origin.

Their forms are best understood by the figures 6, *a*, *b*, *c* and *d*. The examination of the fine slimy residues from the solution of various primary crystalline limestones, in which, from the absence of well marked foreign minerals, it may be difficult to prove the presence of distinct organic forms, will, I think, afford the quickest and readiest mode of establishing the existence of organisms.

The presence of the Eozoon in the primary limestone of Steinhag being thus established, I proceeded to examine such specimens as were at my disposal from other localities of similar limestones in the vicinity of Passau. I must here remark that these specimens, collected during my geological examinations twelve years since, were chosen as containing intermixtures of serpentine and hornblende, and not with reference to the possibility of their holding organic remains. I succeeded however in detecting at least traces of Eozoon in specimens of the limestone from Untersalzbach, (fig. 2,) from Hausbach, Babing, (fig. 3,) and from Kading and Stetting. Moreover a specimen of ophicalcite from a quarry near Srin, in the region between Krumau and Goldenkron, among the primitive hills of Bohemia, afforded unequivocal evidences of Eozoon. Von Hochstetter moreover has received specimens of crystalline limestone from the same strata at Krumau, in which Dr. Carpenter has shown the presence of Eozoon. To the same formation belong the calcareous rocks near Schwarzbach, in the vicinity of which, as near Passau, great masses of graphite are intercalated in the gneiss hills. These limestones of Schwarzbach connect those of Krumau with the similar strata near Passau, from which they are only separated by the great granite mass of the Plockenstein hills. We thus obtain a still farther proof of the similarity of structure throughout the whole range of primitive rocks of Bavaria and Bohemia; and of the parallelism of their lowest portion with the Laurentian gneiss system of Canada. I think therefore that we may, withou hesitation, *place the Hercynian gneiss formation of the mountains forming the Bavarian and Bohemian frontier, on the same geological horizon with the Laurentian system.*

Farther northward, in similar gneiss hills, occupying a limited area, a crystalline limestone occurs near Burggrub, not far from Erbendorf, from which a few specimens were at hand. They were however a reddish, very ferruginous dolomite, penetrated by fibres of hornblende and epidote, and gave me no trace of organic remains.

Besides these limestones of the Hercynian gneiss, there is found in Bavaria another remarkable deposit of crystalline limestone, included in the Hercynian primitive clay-slate series on the south and south-east border of the Fichtelgebirge, in the vicinity of Wunseidel. This clay-slate formation, as we have already shewn, overlies the Hercynian gneiss and mica-slate series, and is immediately beneath the primordial zone of the Lower Silurian strata met with in the Fichtelgebirge. It would thus seem to correspond with the Cambrian rocks of Wales, and with the Huronian system of Canada, as Sir Roderick Murchison has already suggested. This view is confirmed by Fritzsch's discovery of traces of annelids in the grauwacke of Przibram, and by the occurrence of crinoidal stems and foraminiferal forms, according to Reuss, in the limestone of the primitive clay-slates of Paukratz, near Reichenstein. Thus our Hercynian mica-slate, with certain hornblendic strata and chloritic schists belonging to the same horizon, would occupy a stratigraphical position similar to the Labrador series, or Upper Laurentian, of Canada.

The crystalline limestone of the Fichtelgebirge forms in the primitive clay-sl.. two nearly parallel bands, which I conceive to be the outcrops of one and the same stratum, on the opposite sides of a trough. It presents several parallel beds separated by intervening beds of the conformable clay-slate.

The limestone strata near Wunseidel dip from 50° to 75° S.E., and sometimes attain a thickness of 350 feet. They are in many places dolomitic.   *   *   *   *   Spathic iron, in nests and disseminated, characterizes this rock, and by its decomposition gives rise to the valuable deposits of brown hematite, which are worked along the outcrop of the limestone band. Among the other minerals may be mentioned graphite, in crystalline plates, and also in small round grains and rounded compact masses in the limestone; besides which it frequently enters into the composition of the adjacent clay-slate, giving rise to a plumbaginous slate. Fluor-spar, chondrodite, tremolite, common hornblende, serpentine, cubic and magnetic pyrites, are among the minerals of the limestone. Quartz secretions are also met with, but are evidently of secondary origin. The hornblende forms rounded patches, remarkable twisted stripes, and banded parallel layers, often of considerable dimensions, as in the specimens from Wunseidel, which exhibit sheets of hornblende of from five to fifteen millimeters, separated by limestone layers of from fifteen to twenty millimeters in thickness. My examinations of the specimens of this nature, in my collection, have not enabled me to connect these hornblende layers with organic structure, nor to discover any traces of Eozoon in the highly crystalline limestone.

The result of my examinations of specimens of the limestone containing serpentine from the quarries near Wunseidel, from Thiersheim, and from between Hohenberg and the Steinberg, were however more successful. Fragments of the rock from near Hohenberg show irregular greenish stripes, which are made up of parallel undulating laminæ, or of elongated grains. This banded aggregate is a granular mixture of carbonate of lime, serpentine, and a white mineral, insoluble in acids, which appears to be a variety of hornblende. The grains of this aggregate have generally a diameter of $\frac{1}{10}$ millimeter.

When examined in thin sections, the calcareous portions appear for the most part sparry, and traversed by straight intersecting lines, (Pl. 1, fig. 7 a,) or divided into cellular spaces by small irregular bands, which, after the surface is etched, are seen in slight relief. The portions between these bands are granulated. (fig. 7 b.) More compact calcareous portions are however met with, and these are penetrated by delicate tufts of tubuli like those of Eozoon, (fig. 7 c,) and are adherent to the serpentine portions, which have nearly the same form as in the Eozoon of Steinhag, but are far smaller. (fig. 7 d.) In decalcified specimens, they are found to possess the same arched walls as the Eozoon. Their breadth in the cross section is generally about one tenth, and the diameter of the casts of the tubuli only about one hundredth of a millimeter. These broader serpentine portions are generally connected with an adjacent portion of lamellæ, (also composed of serpentine, or of a whitish mineral,) which are not more than one-half their size, curiously curved, and presenting highly arched and deeply incurved outlines, as may be seen in decalcified specimens, (fig. 7 e.) The study of these structures leaves no doubt that they are due to an organism belonging to the same group as the Eozoon. In order however to distinguish this distinctly smaller form of the primitive clay-slate series, with its minute contorted chambers filled with serpentine, from the typical Eozoon Canadense of the more ancient Laurentian system, it may be designated as Eozoon Bavaricum.

I have moreover subjected to microscopic examination a series of specimens from the same limestone horizon in the Fichtelgebirge, which, unlike those   described, showed no distinct foreign minerals, although presenting certain dense portions which seemed to indicate the presence of some foreign matter. These portions however showed only a cellular structure, like that in the specimen from Hohenberg, without any tubuli; nor did etching succeed in developing any structure in these wholly calcareous specimens. When therefore carbonate of lime both constitutes the skeleton, and replaces the sarcode, there is evidently little hope of recognizing these organic forms. If however the flaky pellicles which remain suspended in the acid after the solution of the lime, in these almost wholly calcareous specimens, are examined, they present a very great resemblance to the similar pellicles from the Eozoon limestone of Steinhag, already figured, which have such a striking resemblance to organic forms. The careful examination of the limestone from many other parts in the Fichtelgebirge, affords evidence of organic life similar to those of Hohenberg; thus tending more and more to fill up the interval between the Laurentian gneiss, and the primordial zone of the Lower Silurian fauna. We may therefore reasonably hope that in the study of these more ancient rock-systems, which geologists have only recently ventured to distinguish, paleontological evidence will be found no less available than in the more recent sedimentary formations. The inferences which we are permitted to draw from the discovery of organic remains in these ancient rocks, confirm the conclusion to which I had previously arrived from the study of the stratigraphical relations, and the general character of these ancient rock-systems; viz., *that there exists, in these ancient crystalline stratified rocks, a regular order of progress determined by the same laws which have already been established for the formations hitherto known as fossiliferous.*

I cannot conclude this notice of the preliminary results obtained in the investigation of the ancient Eozoon limestones of Bavaria, without adding a few observations upon some foreign crystalline limestones. It is well known that the crystalline minerals, which in numerous localities are found in these limestones, often present rounded surfaces, as if they had at one time been in a liquid state. As examples of these, Naumann mentions apatite, chondrotite, hornblende, pyroxene, and garnet. The edges and angles of these are often rounded; the planes curved or peculiarly wrinkled, and only rarely presenting crystalline faces; having in short a half-fused aspect, and offering a condition of things hitherto unexplained. One of the best known instances of this is found in tho green hornblende (pargasite) from Pargas in Finland. This mineral there occurs in a crystalline limestone with fluor, apatite, chondrotite, pyroxene, pyrallolite, mica and graphite; associations very similar to those of the serpentine of Steinhag. The graius of pargasite, although completely crystalline within, and having a perfect cleavage, are rounded on the exterior, curved inward and outward, and also approximatively cylindrical in form; so that they may be best compared with certain vegetable tubercles. If the crystalline carbonate of lime which accompanies the pargasite is removed by an acid, there remains a mass of pargasite grains, generally cohering, and presenting a striking resemblance to the skeleton obtained by submitting the Eozoon serpentine-limestone to a similar treatment. The tubercles of pargasite are then seen to be joined together by short cylindrical projections, which are however readily broken by pressure, causing the mass to separate into detached grains. The highly crystalline and ferruginous carbonate of lime which is mingled with the pargasite, shews no organic structure either when etched or examined in thin sections; although the pargasite presents forms similar to those observed in the serpentine of Steinhag. The surfaces of the curved cylindrical and tuberculated grains of pargasite are in part naked, and in part protected by a thin white covering. In some parts fine cylindrical growths are observed, and in others cylindrical perforations passing through the grains of pargasite. By a careful microscopical examination of the surface of these grains (Pl. I., fig. 8), numerous small tubuli, sometimes two millimeters in length, are clearly seen, and by their exactly cylindrical form may be readily distinguished from other pulverulent, fibrous and acicular crystalline mineral matters. These cylinders consist of a white substance, which contrasts with the dark green pargasite, and have the diameter of the tubuli of Eozoon, or from $\frac{1}{1000}$ to $\frac{1}{600}$ millimeters. A single large cylinder was also observed lying obliquely across between two of the pargasite tubercles. (Pl. I., fig. 8 a.) In the decalcified specimens, a white mineral, probably scapolite, was observed side by side with the green pargasite; sometimes forming groups of tubercles like the latter; while in other cases a single tubercle was found to be made in part of the green and partly of the white mineral. From these observations there can scarcely remain a doubt that these curiously rounded grains of pargasite imbedded in the crystalline limestone of Pargas represent the casts of sarcode-chambers, as in the Eozoon; and that they are consequently of organic origin. From the great similarity between the forms of the pargasite grains and the Eozoon-serpentine, we may fairly be permitted to assume the presence of Eozoon in the crystalline limestones of Finland.*

Similar relations are doubtless to be met with throughout the crystalline limestones of Scandinavia, wherever such mineral species occur in rounded grains or in tuberculated forms. The notion that these forms are of organic origin, and have been moulded in the spaces left in a calcareous skeleton by the decay of animal matter, receives a strong support from the observations of Nordenskiold and Bischof. The former found in a tuberculated pyrallolite, 6·38 per cent. of bituminous matter, besides 3·58 per cent. of water; while Bischof states that the same mineral becomes black when ignited, and when calcined in a glass tube, gives off a clear water with a very offensive empyreumatic odor.

There may also be mentioned in this connection a phenomenon which is probably related to those just described. Upon the pyritous layers which occur in the Hercynian gneiss near Boden, are found great quantities of grains of quartz, almost transparent, and with a fatty lustre, which have in all cases rounded undulating forms, precisely resembling the pargasite tubercles from Finland. Dichroite also sometimes occurs in this region in similar shapes, although it also, in many cases, forms perfect crystals. The evidence of organic forms may perhaps be found in these masses of quartz and dichroite, though their treatment will necessarily present difficulties.

A specimen of crystalline limestone, with rounded pyroxene (coccolite) grains from New York, showed, after etching by means of acids, no traces of tubuli; but the grains of coccolite, remaining after the entire removal of the carbonate of lime, were found to be connected with each other by numerous fine cylindrical tubuli and skin-like laminæ. The surface of the rounded coccolite grains was much wrinkled, and studded with small cylindrical processes of a white mineral, sometimes ramifying, and apparently representing the remnants of a system of tubuli which had been destroyed by the crystallization of the carbonate of lime. The flaky residue from the solvent action of the acid exhibits, under the microscope, laminæ, needles, and strings of

* These belong to the primitive gneiss formation of Scandinavia, which the geologists of Canada, so long ago as 1855, referred to the Laurentian system.—T. S. H.

globules similar to those described in the residue from the Eozoon ophicalcite of Steinhag, with which, and with the hornblendic limestone of Pargas, this coccolite-bearing limestone of New York seems to be closely related.

A fragment of ophicalcite from Tunaberg in Sweden bears a striking resemblance to the coarser marked varieties of this rock from near Passau. The carbonate of lime between the tubuli is very sparry; and after its removal, a perfectly coherent serpentine skeleton is obtained, as in the Passau specimens. The surface of the serpentine tubercles is abundantly covered with acicular crystalline needles of various lengths, whose inorganic nature is unmistakeable. The sediment from the acid solution also contains a prodigious quantity of these same small crystalline needles. On etching a specimen of this rock with dilute acid, the same needles were found in most places; but here and there, in isolated, less crystalline and more solid portions of the carbonate of lime, there were seen curved and ramified tubuli, undoubtedly corresponding with the tubuli of Eozoon, and having the same size and manner of grouping as in the Eozoon of Passau. The ophicalcite of Tunaberg is therefore to be classed with the Eozoon-bearing limestones.

A specimen of crystalline limestone from Boden in Saxony, holding rounded grains of chondrodite, hornblende and garnet, and furnished me by Prof. Sandberger, showed, after etching, tubuli of surprising beauty, both singly and in groups, but only in small isolated compact portions of the carbonate of lime. The sparry crystallization of this mineral seems to have frequently destroyed the cohesion of the very delicate tubuli, the fragments of which may be observed in very large quantity in the flaky residue from the solution.

A blackish serpentine limestone from Hodrisch in Hungary, showed by etching no traces of tubuli. The granular residue from its solution in acids showed under the microscope large quantities of cell-like granules, with a central nucleus, and generally joined in pairs, like the spores of certain lichens. More rarely however three or four of such grains were joined together. By far the greater part of them were of one and the same size, although occasionally others of double size were met with. Their regularity of form is much in favor of their origin from organic structure.

A fragment of ophicalcite from Reichenbach in Silesia, which Prof. Beyrich kindly furnished me, showed distinct parallel bands of serpentine with curved and undulating outlines, resembling the Eozoon ophicalcite of Canada. The etched portions show, in the carbonate of lime between the serpentine, or in the interspaces of the serpentine, the same relations as the limestone of Hohenberg from the primitive clay-slate formation. The tubuli, which have a certain resemblance with those of Hohenberg, are stuck together, as if covered by an incrustation. Further examinations of this limestone are required to determine more definitely the organic nature of its enclosures.

A fragment of similar limestone without serpentine, from Raspenau, shows not the remotest trace of any organic structure whatever. The same negative results were obtained with a specimen of granular limestone from Timpobepa in Brazil; and with a very coarsely crystalline carbonate of lime, holding chondrodite, from Amity, New Jersey. These negative results show that organic remains are sometimes wanting in the primitive crystalline limestones, as well as in those of more recent formations. The occasional absence from the primary limestones of these regular structures is therefore an indirect argument for their organic origin.

Explanation of the Plate.

Figure 1. Section of *Eozoon Canadense*, with its serpentine replacement, showing the fine tubuli and the canal-system, from the limestone of the Hercynian gneiss formation at Steinhag; seen by reflected light, and magnified 25 diameters.

2. Section of Eozoon from the limestone of Untersalzbach; 25 diameters.

3. Section of Eozoon from the limestone of Bubing.

4. Section of Eozoon from the limestone of Steinhag; 120 diameters.

5, *a* and *b*. Knotted tubuli from the insoluble residue of the Steinhag limestone; 300 diameters.

6, *a*, *b*, *c*, and *d*. Flocculi from the same residue; 400 diameters.

7. Section of *Eozoon Bavaricum*, with serpentine, from the crystalline limestone of the Hercynian primitive clay-state formation at Hohenberg; 25 diameters.

   a. Sparry carbonate of lime.

   b. Cellular carbonate of lime.

   c. System of tubuli.

   d. Serpentine replacing the coarser ordinary variety.

   e. Serpentine, and hornblende, replacing the finer variety, in the very much contorted portions.

8. Aggregated grains of pargasite, remaining after the solution of the carbonate of lime, from the granular limestone rock of Pargas.